U.S. Department
of Transportation
**National Highway
Traffic Safety
Administration**

DOT HS 811 115

April 2009

The Effectiveness of Amber Rear Turn Signals for Reducing Rear Impacts

This report is free of charge from the NHTSA Web site at www.nhtsa.dot.gov

This publication is distributed by the U.S. Department of Transportation, National Highway Traffic Safety Administration, in the interest of information exchange. The opinions, findings, and conclusions expressed in this publication are those of the author and not necessarily those of the Department of Transportation or the National Highway Traffic Safety Administration. The United States Government assumes no liability for its content or use thereof. If trade or manufacturer's names or products are mentioned, it is because they are considered essential to the object of the publication and should not be construed as an endorsement. The United States Government does not endorse products or manufacturers.

Technical Report Documentation Page

1. Report No. DOT HS 811 115	2. Government Accession No.	3. Recipient's Catalog No.
4. Title and Subtitle The Effectiveness of Amber Rear Turn Signals for Reducing Rear Impacts		5. Report Date April 2009
		6. Performing Organization Code
7. Author(s) Kirk Allen, Ph.D.		8. Performing Organization Report No.
9. Performing Organization Name and Address Evaluation Division; National Center for Statistics and Analysis National Highway Traffic Safety Administration Washington, DC 20590		10. Work Unit No. (TRAIS)
		11. Contract or Grant No.
12. Sponsoring Agency Name and Address National Highway Traffic Safety Administration 1200 New Jersey Avenue SE. Washington, DC 20590		13. Type of Report and Period Covered NHTSA Technical Report
		14. Sponsoring Agency Code
15. Supplementary Notes		
16. Abstract This purpose of this report is to determine the effect of rear turn signal color on the likelihood of being involved in a rear-end crash. Federal Motor Vehicle Safety Standard No. 108 allows rear turn signals to be either red or amber in color. Previous work on this subject includes laboratory experiments and analyses of crash data that suggest amber rear turn signals are beneficial. The present study was designed around the concept of "switch pairs" – make-models of passenger vehicles were identified that had switched rear turn signal color, and crash involvement rates were computed before and after the switch. This method should control for extraneous factors related to vehicle and driver characteristics. Crash data from NHTSA's State Data System was used in the analysis. The principal finding of the report is that amber signals show a 5.3% effectiveness in reducing involvement in two-vehicle crashes where a lead vehicle is rear-struck in the act of turning left, turning right, merging into traffic, changing lanes, or entering/leaving a parking space. The advantage of amber rear turn signals is shown to be statistically significant.		
17. Key Words NHTSA; NCSA; State Data System; crash avoidance; automotive lighting; turn signal color		18. Distribution Statement This report is free of charge from the NHTSA Web site at www.nhtsa.dot.gov
19. Security Classif. (Of this report) Unclassified	20. Security Classif. (Of this page) Unclassified	21. No. of Pages 44	22. Price

Form DOT F 1700.7 (8-72)   Reproduction of completed page authorized

Table of Contents

List of Figures ii
List of Tables iii

Executive Summary iv

Introduction 1
    Previous Research 1
    Effectiveness of other conspicuity and lighting systems 3
    Discussion 3

Methods 4
    Crash Scenario 4
    Control Group 4
    Turn Signal Color Identification 5
    Analytical Approach 6
    Data Sources 9

Results 10
    Stage One 10
    Stage Two 10
    Stage Three 11
    Stage Four 11

Supplementary Analyses 12
    Injury Outcome 12
    Vehicle Maneuver 13
    Gender 15
    Vehicle Age 15
    Observed effectiveness, State by State 18
    Alternate Control Group 19

Appendix 21
    Methods: Turn Signal Color Identification 21
    Methods: Data Sources 24
    Methods: State Data Summary 25
    Results: Stage One 30
    Results: Stage Four 31
    Supplementary Analyses: Injury Outcome 35

List of Figures

Figure 1: Make-Models and Model Years included in study 6
Figure 2: Odds ratio by vehicle age, with 95 percent one-sided confidence intervals (reference marked for overall odds ratio) 18
Figure 3: Odds ratio by State, with 95 percent one-sided confidence intervals (reference marked for 1.00) 19
Figure 4: Nissan Altima, MY 1998-1999, amber turn signals 21
Figure 5: Nissan Altima, MY 2000-2001, red turn signals 22
Figure 6: Nissan Altima, MY 2002-2004, amber turn signals 22
Figure 7: Nissan Altima, MY 2005, red turn signals 23
Figure 8: States and calendar years of State data included in study 24
Figure 9: Number of struck vehicles, by vehicle maneuver for 14 States 26
Figure 10: Relative frequency of struck vehicle maneuvers, by State 27
Figure 11: Number of struck vehicles, for the 26 make-models that switched signal color 28
Figure 12: Relative frequency of struck vehicle maneuvers, by make-model 29
Figure 13: Chevrolet Camaro, MY 1991-1992, amber 32
Figure 14: Chevrolet Camaro, MY 1993-1994, red 32
Figure 15: Chevrolet Camaro, MY 1995-1996, red 33
Figure 16: Chevrolet Camaro, MY 1997-1998, amber 33

List of Tables

Table 1: Hypothetical 2×2 table for illustrative purposes..7
Table 2: Hypothetical example for those who do not use turn signals 9
Table 3: Hypothetical example for those who do use turn signals 9
Table 4: 2×2 table excluding major redesigns...11
Table 5: 2×2 table excluding minor and major redesigns..12
Table 6: 2×2 table for crashes with injuries..13
Table 7: 2×2 table for crashes without injuries...13
Table 8: 2×2 table for crashes where the struck vehicle was changing lanes or merging 14
Table 9: 2×2 table for crashes where the struck vehicle was changing lanes, merging, parking, or making a U-turn..14
Table 10: 2×2 table for crashes where the struck vehicle was turning right....................14
Table 11: 2×2 table for crashes where the struck vehicle was turning left......................14
Table 12: 2×2 table for crashes where the driver was male...15
Table 13: 2×2 table for crashes where the driver was female..15
Table 14: Number of struck vehicles by age at time of crash...17
Table 15: 2×2 table using non-turning rear-struck as the control group..........................20
Table 16: Summary of types of crashes, by State..25
Table 17: Counts of collision types for struck vehicles, by State....................................27
Table 18: Counts of collision types for struck vehicles, by make-model........................29
Table 19: Struck and striking cases relative to overall number of crashes, by State 30
Table 20: List of effectiveness for 33 make-models...31
Table 21: Make-models that switched from red to amber (13) 34
Table 22: Make-models that switched from amber to red (20) 35
Table 23: Crash severity for Florida..36
Table 24: Classification of crashes by injury, struck vehicles only.................................36

Executive Summary

This report analyzed the effect of rear turn signal color as a means to reduce the frequency of passenger vehicles crashes. Specifically, an answer was sought as to whether amber or red turn signals were more effective at preventing front-to-rear collisions when the rear-struck (leading) vehicle was engaged in a maneuver where turn signals were assumed to be engaged – turning, changing lanes, merging, or parking.

Using NHTSA's State Data System (SDS), the process was two-fold: (1) identify crashes with assumed turn signal engagement, restricted to crashes involving two passenger vehicles; and (2) identify make-models that switched signal color from amber to red, or vice versa, at some point on the range of 1981 to 2005. These two steps created a dataset of front-to-rear collisions involving make-models that had switched signal color.

Data from 14 States contained the necessary variables to identify the relevant crashes and characteristics of the vehicles involved. Twenty-six make-models were identified that had changed signal color, a total of 33 switches counting several that changed more than once. Under these conditions, it would likely require great effort to arrive at notably larger sample sizes. Despite this breadth, the dataset is not without limitations. For example, only one western State (Utah) is included and there are few LTVs. It is also not possible to evaluate the latest model years, which have seen some changes in vehicle lighting such as LEDs.

Several stages of analysis were conducted. As per the ability to prevent involvement in crashes with assumed turn signal engagement, amber turn signals proved significantly more effective at each stage, compared to red turn signals. The single best point estimate of the effectiveness is 5.3 percent, based on a pool of make-models that are as nearly similar as possible aside from the color of the rear turn signals, i.e., they did not change body size, body style, or the size and shape of the rear lighting housings. The result is significantly different from zero ($\chi^2 = 5.17$, with 1 df, $p < 0.02$), with a 95-percent one-sided confidence interval indicating the effectiveness is at least 1.5 percent.

The magnitude of the estimate (5.3%) is similar to the long-term magnitude of the CHMSL effectiveness (4.3%), but this is smaller than reported in cited analyses of rear turn signals and other automotive lighting features (~20%, refer to Background section). Compared to two other published analyses of amber turn signals, the "switch pair" method in this report should provide greater control of extraneous factors that influence crash involvement. Although the magnitude of the effectiveness is lower, this study is consistent with the cited reports in finding amber turn signals significantly more effective than red turn signals.

In several supplementary analyses, it was suggested that amber turn signals may be *slightly* more effective in crashes involving injuries (Table 6), changing lanes or merging (Table 8), and left turns (Table 11). However, these results are *not sufficiently strong to be considered different* from the primary result of 5.3 percent effectiveness.

Introduction

Rear turn signals on passenger vehicles in the United States are permitted to be either amber or red, according to Federal Motor Vehicle Safety Standard (FMVSS) No. 108. Previous research has shown amber rear turn signals to be beneficial at reducing the likelihood of being rear-ended while engaging in maneuvers that involve signaling.

This report seeks to verify and quantify the crash-reduction benefits of amber rear turn signals using real-world crash data from NHTSA's State Data System. The intent is to draw a generality from a large amount of existing data.

In Europe and many other countries, rear turn signals are required to be amber. Currently, manufacturers who produce vehicles for both the European and North American markets choose whether to produce all vehicles with amber rear turn signals or to equip the North American products with red rear turn signals. That is to say, the European regulations are more restrictive rather than being in conflict with FMVSS No. 108.

Previous Research

Several laboratory experiments were conducted in the period 1968 to 1977,[1] measuring the time subjects took to respond to a brake light on a lead vehicle. Several combinations of rear lighting parameters and crash scenarios were evaluated, e.g., separated versus combined turn/tail/stop lamps, red versus amber signal color in separated housings, turning while stopping versus turning while not stopping. Viewed as a whole, these studies indicate faster subject response time when the rear lighting has a greater degree of separation and is multicolored. These studies establish a theoretical basis for amber rear turn signals (in combination with red brake lamps and tail running lamps) being more beneficial than all-red signals and lamps.

In a simulated driving task, Luoma and colleagues[2] assessed the effect that turn signals have on recognition of brake signals. The dependent measure is the time to react to a brake signal depending on whether the turn signal is illuminated. The between-subject independent variables were signal lamp color (amber, red); lamp condition (brake signal only, brake signal preceded by turn signal); and luminous intensity of the turn signal (80 cd, 130 cd). The within-subject variables were age and gender. Three main effects were statistically significant – color (red turn signals had longer response times), lamp condition (reaction times were longer with the turn signal on), and age (older participants took longer to respond). Averaged across luminous intensity and signal condition, amber turn signals yielded response times about 20 percent faster compared to red turn signals. That is to say, subjects responded more quickly to the brake signal when the turn signal was amber. The number of errors was assessed based on falsely responding to a brake

---

[1] Most of these were conducted by Mortimer and are summarized in Edwards' paper, discussed below (footnote no. 3).

[2] Luoma, J., Flannagan, M. J., Sivak, M., Aoki, M., & Traube, E. C. (1995, February). "Effects of turn-signal color on reaction times to brake signals," UMTRI-95-5; also published in *Ergonomics*, 1997, vol. 40 no. 1, pp. 62-68

signal when only the turn signal was active or taking excessively long to respond when the brake signal was active. Amber signals were more effective than red at reducing the likelihood of respondents making an error of these types.

Analyses of real-world crash data are based on identifying two types of crashes. The first type are those crashes where turn signal use is likely to be influential, e.g., turning left, turning right, merging into traffic, and changing lanes. Second, a control group is constructed based on some measure of exposure. Several examples follow.

Edwards[3] analyzed crashes in five States occurring between July 1, 1983, and June 30, 1985. He found that amber turn signals were 20.4-percent effective at reducing rear-end crashes with "total rear-end crashes" as the control group of crashes and 17.7-percent effective with "total crashes" as the control group. For both control groups, vehicles equipped with amber rear turn signals were favored in all five States in the study. A supplementary control group was constructed based on vehicle registration numbers from these States, and amber once again won out in all five.

Taylor and Ng[4] analyzed insurance claims in Canada. Make-models were included on the range of model years 1975 to 1979 that had switched from a non-separated all-red rear lighting configuration to a separated system with amber turn signals. To control for vehicle characteristics, make-models were included that switched configurations in consecutive models years or else were the same car body under different marques in a given model year. The rate of being struck in the rear while turning was compared to a control group of being struck in the rear while not turning. The results did not show a significant effect in favor of either rear lighting configuration. The study is limited by a small sample size (400 rear-struck turning cases) and a vehicle pool that may not be representative of the overall on-road fleet at that time. The results could be confounded by turn signal separation because all red systems were un-separated and by vehicle age because all changes were *to* the amber/separated configuration.

Sullivan and Flannagan[5] conducted logistic regression of crash data. They first catalogued rear lighting characteristics of the 50 most common make-models from calendar year 2003. Along with turn signal color, these characteristics included turn signal lens opacity, turn signal light source, turn signal optics, and rear signal separation. Using NHTSA's State Data System, these make-models were then identified in crash involvements where a vehicle is struck in the rear while turning, merging, or changing lanes. Among four analyses, the most tightly controlled is restricted to make-models that

---

[3] Edwards, M. L., (1988). "An Investigation of Selected Vehicle Design Characteristics Using the Crash Avoidance Datafile,," *Eleventh International TechnicalConference on Experimental Safety Vehicles*. NHTSA Report No. DOT HS 807 223 pp. 389-395. Washington, DC: National Highway Traffic Safety Administration.

[4] Taylor, G. W., and Ng., W. K. (1981). "Effectiveness of Rear-Turn-Signal Systems in Reducing Vehicle Accidents from an Analysis of Actual Accident Data," *SAE Technical Paper Series*, no. 810192, Detroit, 1981.

[5] Sullivan, J. M., & Flannagan, M. J. (2008). "The Influence of Rear Turn Signal Characteristics on Crash Risk," NHTSA Report No. DOT HS 811 037. Washington, DC: National Highway Traffic Safety Administration.

switched colors during the model years 1990 to 2005. The primary result is a 22-percent crash reduction for vehicles with amber turn signals (95% confidence interval 12% to 30%). If lamp separation is placed in the regression model in place of turn signal color, separation is itself a significant predictor, although weaker than signal color.

Effectiveness of other conspicuity and lighting systems

Earlier studies consistently point to a crash-reduction advantage for amber turn signals. For comparison, the center high mounted stop lamps[6] (CHMSL) were estimated to reduce rear impact crashes by 8.7 percent in the first model year after being mandated (1987 for passenger cars), and the long-term effectiveness was 4.3 percent (postulated to have decreased due to driver acclimatization to the devices).

Side marker lamps became required for large trucks and buses on January 1, 1968, and on January 1, 1969, for all other vehicles. These small lamps are illuminated when the headlights are on and serve as reflectors at other times. The side marker lamps enhance conspicuity such that the would-be-impacting vehicle can take evasive maneuvers. Prior to their introduction, vehicles had no side illumination. A NHTSA analysis[7] found that side marker lamps result in a 16-percent reduction in nighttime, side-angle collisions. The effectiveness was slightly higher in accidents with personal injury, at 21 percent, although no fatality-reduction benefit was evident.

Red-and-white retroreflective tape is required on all heavy trailers manufactured after December 1, 1993. Older trailers were to be retrofitted with some type of conspicuity treatment by June 1, 2001, thus nearly every on-road trailer had some conspicuity enhancement by that time. The purpose of retroreflective tape is similar to side marker lamps – to enhance the visibility of vehicles that might not otherwise be seen at night, thus preventing side and rear impacts. An analysis by NHTSA[8] reported a 29-percent crash reduction benefit. The effect was greatest under "dark-not-lighted" conditions (41%) but was not statistically significant in other conditions ("dark-lighted," "dawn," "dusk"). The tape was most effective (44%) at preventing crashes with at least one injury or fatality.

Discussion

Two analyses of crash data report that amber turn signals are approximately 20 percent more effective than red turn signals at preventing vehicles from being rear-ended while performing a maneuver that typically involves signaling. One study did not find a statistically significant difference between red and amber turn signals. There has been no research presented that is in favor of rear turn signals being red.

---

[6] Kahane, C. J., & Hertz, E. (1998). "The Long-Term Effectiveness of Center High-Mounted Stop Lamps in Passenger Cars and Light Trucks. NHTSA Report No. DOT HS 808 696. Washington, DC: National Highway Traffic Safety Administration.

[7] Kahane, C. J. (1983). *An Evaluation of Side Marker Lamps for Cars, Trucks, and Buses*. NHTSA Report No. DOT HS 806 430. Washington, DC: National Highway Traffic Safety Administration.

[8] Morgan, C. (2001). *The Effectiveness of Retroreflective Tape on Heavy Trailers*. NHTSA Report No. DOT HS 809 222. Washington, DC: National Highway Traffic Safety Administration.

Analyses of three other conspicuity devices found crash reductions of 4.3 percent (CHMSL), 21 percent (side marker lamps), and 29 percent (heavy trailer retroreflective tape). These latter three were *additions* to vehicles, whereas amber turn signals are a modification (change in color) to an *existing* technology (rear turn signals). This information sets some expectations for the magnitude of the effectiveness of amber turn signals.

Methods

The preceding section on Previous Research suggests several guiding principles. First, crashes are identified where the safety device is expected to have a benefit. Second, some contrast group of crashes serves as a control group – a measure of overall crash exposure in situations where the safety device should have no influence. Third is the concept of comparing crash involvement rates before and after the introduction of a safety device.

Crash Scenario

It is proposed that rear turn signal color is influential in crashes where a careful driver usually activates the turn signals – turning left, turning right, changing lanes, merging into traffic, exiting from traffic, making a U-turn, entering parking, or leaving parking. To identify crashes where rear turn signals are most conspicuous, the following restrictions are placed on the crash:
- It is a two-vehicle crash, in which both vehicles are passenger vehicles – cars or LTVs.
- One vehicle is struck in the rear, and the other is struck in the front – that is, the other does the *striking* with its front.
- The maneuver of the striking vehicle is not taken into consideration – the struck vehicle defines the crashes of interest.

Control Group

Ideally, crash involvement while signaling could be compared to crash involvement while not signaling, in a set of crashes with similar circumstances. Because signal activation is not collected in any existing database, it is necessary to compare the involvement in crashes where the signals are assumed to be engaged to some other type of crash where signals would not normally be engaged.

On the surface, it seems appropriate to identify a control group as those vehicles rear-struck in situations when the turn signals are not engaged, i.e., being rear-struck in a maneuver other than turning, changing lanes, merging, or parking. Unfortunately, one cannot be certain that the struck vehicles are *not* signaling. For example, a driver may be approaching an intersection with the intention to turn and have his signal activated but not yet be engaged in a maneuver that a police report would classify as "turning.." Because there are so many ways to be rear-struck, there may be variability in crash classification, according to policies and procedures of police reporting at the State or Local level.

A control group of striking vehicles is defined as the vehicles that strike with their front portions into vehicles engaged in maneuvers where the turn signal is assumed to be activated. The dataset is restricted to crashes with exactly two passenger vehicles, both of which are identifiable in the crash data files. Therefore, the number of struck (leading) and striking (trailing) vehicles is identical.

Turn Signal Color Identification

Turn signal color is not recorded in crash data files. Rear turn signal color was primarily determined from searching auto parts and used car sales Web sites. Images were catalogued and verified across multiple model years. The oldest make-models were obtained from in-house NHTSA lists and are now too old to be verified photographically.

Vehicles included for study were only those that switched signal color, either from *red to amber* **or** *amber to red*. A few make-models were found that switched multiple times, e.g., *amber to red to amber*. In addition to the two model years comprising the switch, one additional model year was included on each side. Thus, for each make-model, the model years were restricted to two-year blocks of each color. A make-model that switched from amber to red, for example, would include two years of amber, followed by two years of red – four consecutive model years in total. Two-year blocks were chosen as a balance between accumulating a large number of cases and maintaining similarity in vehicle and driver characteristics across model years. An example is shown in the Appendix of the Nissan Altima (Figure 4 to Figure 7 in the Appendix).

This "switch pair" concept is designed to control for driver and vehicle characteristics. If not restricted to switch pairs, variability across make-model may introduce confounding factors that could muddle the interpretation of statistical analyses. Using switch pairs, certain factors can be checked to ensure they do not unduly influence the results, e.g., vehicle age and driver gender. Other factors, however, must be assumed to be equal *within* make-model. These factors include the following:
– Turn signal use;
– Turn signal reliability (i.e., burned-out bulbs); and
– Modifications to the original equipment manufacturer rear lighting housings.

So long as these factors do not vary systematically according to signal color, the analyses should be valid. That is to say, for example, there is no reason why the driver of a Ford Ranger with amber turn signals would be more or less likely to activate the turn signals than the driver of a Ford Ranger with red turn signals.

Figure 1 shows the make-models and model-years that are included in the study. The color coding identifies the rear turn signal color.

Figure 1: Make-Models and Model Years included in study

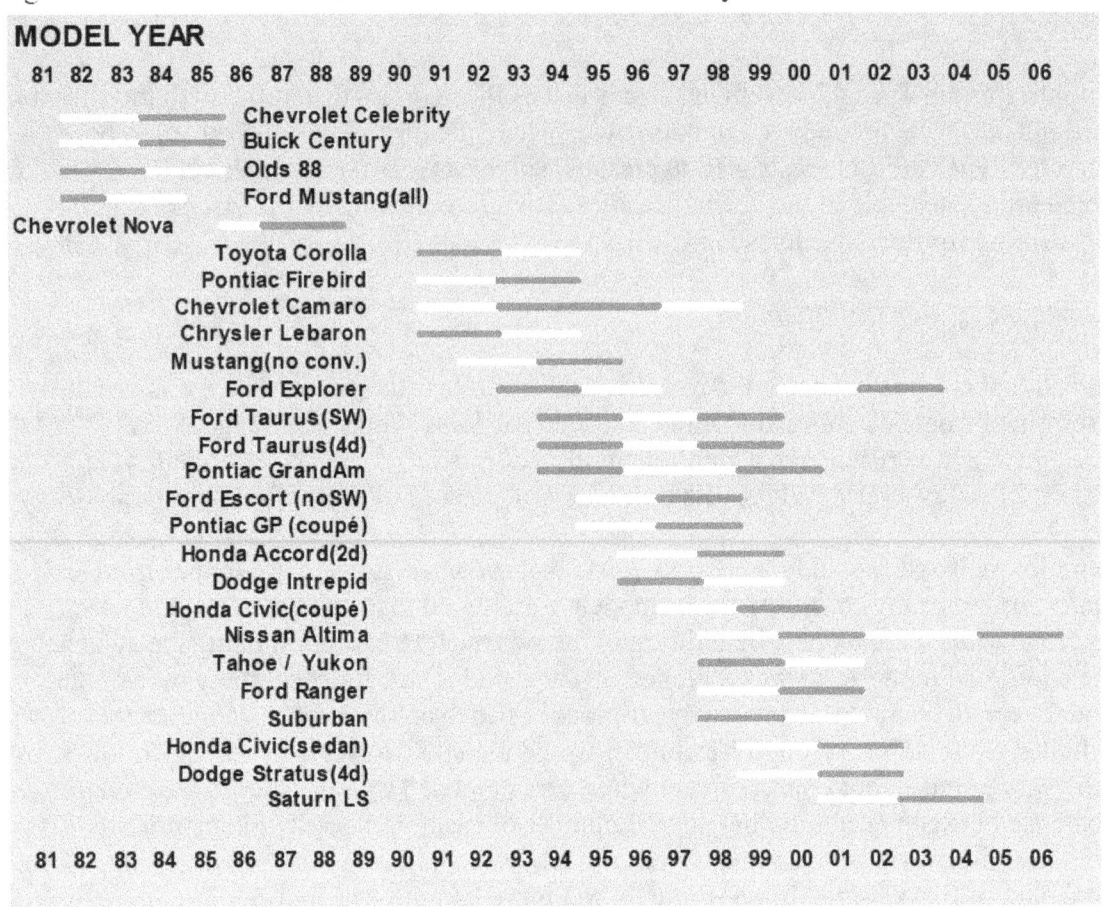

The eligible model years for some make-models are restricted, for several reasons:
- Make-models are too old – VIN identification became accurate after the introduction of the 17-character VIN in MY 1981.
- Make-models are too new – State data files are available up to at most calendar year 2005, except Michigan up to 2006.
- Model years overlap the introduction of CHMSL (beginning MY 1986 in cars and MY 1994 in LTV), which is itself a rear-lighting change of substantial effect.
- It was not possible to verify the signal color with images or from other reports.
- Rear turn signal color varies based on options packages for a given MY.

Analytical Approach

The basic analysis approach is to construct a 2×2 table, such as seen in Table 1. The rows identify the rear turn signal color, and the columns identify the crash involvements. In this hypothetical example, vehicles with red turn signals were the striking vehicle in 100 crashes, compared to being the struck vehicle in 110 crashes. Similarly, there were 150 instances where vehicles with amber turn signals were the striking vehicle, compared to 135 cases where the rear-struck vehicle had amber turn signals.

Table 1: Hypothetical 2×2 table for illustrative purposes

	Striking	Struck	Odds
Red	100	110	1.100
Amber	150	135	0.900
			0.818 Odds Ratio
χ^2 1.22			**18.2% Effectiveness**
p value 0.14		(not significant)	one-sided CI bound

The odds are calculated as the ratio of the number of struck cases to the number of striking cases. For vehicles with red turn signal in this example, the odds of being struck are 1.10, i.e., 110 ÷ 100. The analogous calculation for amber turn signal vehicles yields a value of 0.90. Because the odds of being struck for the amber-equipped vehicles is lower than that for the red-equipped vehicles, it is implied that the amber turn signals were more effective than the red turn signals at preventing rear-impact collisions.

This difference in the odds can be re-expressed as "effectiveness," in this case in favor of the vehicles with amber signals. The odds ratio is calculated by dividing the odds for amber by the odds for red (0.90 ÷ 1.10 = 0.82). The odds ratio is subtracted from one to yield a point estimate of the effectiveness (1 – 0.82 = 0.18 = 18%, favoring amber because it is positive).

The effectiveness can also be interpreted as the extent that the amber vehicles were struck less (in this case) than would be expected if the odds ratio for the red vehicles persisted (i.e., odds ratio for red × number of amber striking vehicles = 1.10 × 150 = 165; and 135 is 18 percent less than 165). An odds ratio of exactly one would yield an effectiveness of 0 percent – no difference between red and amber turn signals. The formulas are defined such that an odds ratio of less than one represents a reduction in crash involvement for amber relative to red. The relevant equations are depicted below.

	Column 1	Column 2
Row 1	A	B
Row 2	C	D

$$\text{Odds(Row1)} = B \div A = \frac{B}{A} \qquad \text{Odds(Row2)} = D \div C = \frac{D}{C}$$

$$\text{Odds Ratio} = \text{Odds(Row2)} \div \text{Odds(Row1)} = \frac{D/C}{B/A} = \frac{D \times A}{C \times B}$$

The statistical significance of the relationships in a 2×2 table is assessed by the chi-square (χ^2). The null hypothesis of the statistical test is that the relative counts in the 2×2 table are equal across rows (the distinction between rows and columns is arbitrary). When expressed as an odds ratio, the χ^2 answers the question of whether the odds ratio is different from one. The χ^2 equals zero when the odds ratio is one and is positive when the odds ratio differs from one. A sufficiently large χ^2 leads to rejection of the null hypothesis, meaning the odds ratio is significantly different from one. The equation to calculate the χ^2 for a 2×2 table is shown below.

$$\chi^2 = \frac{(AD-BC)^2(A+B+C+D)}{(A+B)(C+D)(A+C)(B+D)}$$

degrees of freedom, $df = (rows - 1) \times (columns - 1) = (2 - 1) \times (2 - 1) = 1$

The statistical significance of the relationship in the 2×2 table can be expressed as a confidence interval. Throughout the report, the convention of a 95 percent one-sided confidence interval is adopted. This decision is based on the previous research that indicates amber rear turn signals are superior, i.e., there appears to be no argument explicitly *in favor* of red. For the hypothetical data in Table 1, the one-sided confidence interval on the effectiveness extends to -16.9 percent. When the relationship in a 2×2 table is statistically significant, the effectiveness is expressed as being "greater than or equal to" a certain positive value, e.g., "≥ 2.9%." This sets a lower limit on the effectiveness, meaning that one can be reasonably certain that the true effectiveness is at least equal to that value.

The χ^2 is dependent on both the effectiveness and the sample size – for a given odds ratio, the χ^2 increases as the sample size increases. In Table 1, the χ^2 of 1.22 ($df=1$) is not sufficiently large to conclude that the effect in favor of amber (18.2%) is statistically significant (p-value = 0.14 > 0.05)[9]. To achieve significance at the 0.05 level, the sample size would need to be more than three times larger than in Table 1 (i.e., over 1,500 vehicles if the relative percentages persist). For a 5 percent effectiveness to be statistically significant, a total sample size of nearly 20,000 would be required in the 2×2 table.

The magnitude of the effectiveness is constrained by the frequency of turn signal use. The hypothetical example in Table 1 can be partitioned into those who use the turn signals and those who do not. Table 2 takes a basis of 20 percent non-signaling, applied to the striking vehicles (e.g., for red, 0.20 × 100 = 20). The odds of 1.100 for the red vehicles is held constant from Table 1 and used to perform the other calculations in Table 2. The calculation of effectiveness is exactly zero, reflecting that turn signal color has no effect when the signals are not activated. These 20 percent non-signaling drivers are subtracted from the original data in Table 1 and presented in Table 3. This shows the effectiveness calculation for those who did use the turn signals to be 22.7 percent, compared to 18.2 percent originally.

---

[9] Because a one-sided significance test is used, the reported *p-value* has been divided by 2 compared to a two-sided test.

Table 2: Hypothetical example for those who do not use turn signals

	Striking	Struck	Odds
Red	20	22	1.100
Amber	30	33	1.100

1.000 Odds Ratio

χ^2 0.00 **0.0% Effectiveness**
p value 0.50 (not significant) one-sided CI bound

Table 3: Hypothetical example for those who do use turn signals

	Striking	Struck	Odds
Red	80	88	1.100
Amber	120	102	0.850

0.773 Odds Ratio

χ^2 1.59 **22.7% Effectiveness**
p value 0.10 (not significant) one-sided CI bound

Data Sources

NHTSA's State Data System is the data source for this project. This resource is compiled from police-reported crashes in a State. Thirty-two States have participated since inception of SDS in 1989, but not all years' data have been provided. The type and amount of data collected vary by State according to what is contained on a police report. NHTSA converts the reports to a common format to simplify analyses.

To be eligible for inclusion in this analysis, the following criteria must be met:
- The vehicle maneuver identifies crashes in which the turn signals are assumed to be active. This information is usually coded as VEH_MAN or COL_TYPE – "vehicle maneuver" or "collision type" – depending on the State.
- The vehicle impact location determines whether a vehicle was struck or striking. The struck vehicle was impacted in the rear section – right rear, left rear, or center rear. Similarly, the striking vehicle had an impact in the front section – right front, left front, or center front. These definitions are generally consistent by State. Collisions in which either vehicle was struck in the side are excluded. Both crash-involved vehicles must be listed in the vehicle file, such that the number of *striking* vehicles equals the number of *struck* vehicles (this does not exactly hold when restricted to the relevant make-models, listed in Figure 1).
- Make-model must be identifiable from the VIN or else coded directly into the State file. Texas and Indiana meet this latter criterion, although the full assortment of relevant make-models was not identifiable. The turn signal color is not contained in the State Data files and is gathered from other resources (see page 5), then linked to the crash data according to make-model and model year.

A summary of the States eligible for inclusion in the analysis is shown in Figure 8 in the Appendix (page 24). Following that are several figures and tables describing the types of crashes and the frequencies.

Results

Stage One

Figure 1 identifies 26 make-models that switched rear turn signal color. Including several make-models that switched more than once, there are 33 switch pairs – 20 switched from amber to red, and 13 switched from red to amber. The basic 2×2 table, illustrated in Table 1, is constructed separately for each switch pair. Each table is classified as "favoring amber" or "favoring red." This test should provide maximal control of vehicle and driver characteristics, assuming these factors have greater variability between make-model than within make-model over several years.

For the 33 switch pairs, 24 of these favored amber (24 of 33 = 73%). This result is significantly more than 50 percent, according to a binomial probability test ($p < 0.01$)[10]. This means that it is unlikely to have such a high proportion of make-models favoring amber, compared to an assumption that amber and red are equally effective. Sixteen of the 20 of the amber-to-red switches favored amber, compared to 8 of the 13 red-to-amber switches.

The effectiveness estimates for the 33 switch pairs are shown in Table 22 in the Appendix (page 35).

Stage Two

From the 33 individual effectiveness estimates of Stage One, a parametric estimate of the overall effectiveness was calculated.

The arithmetic mean of the 33 effectiveness estimates is 4.9 percent, and the median effectiveness estimate is 8.4 percent. In principle, the median should be a more accurate point estimate because it is less influenced by extreme values, which arise for a few make-models, partly due to the small sample sizes at this level of detail.

The statistical significance of the mean effectiveness was assessed by first converting each *odds ratio* value to the *log(odds ratio)*. This conversion is necessary because the *odds ratio* values are not normally distributed.[11] The mean of the 33 *log(odds ratio)* is significantly less than zero (Student's $t = -2.15$, $p = 0.04$), meaning in favor of amber. The limit of the 95 percent one-sided confidence interval, re-converted to effectiveness, is 1.4 percent.

---

[10] Binomial probability calculators are readily available on the Web. In Microsoft Excel, the *BINOMDIST* function can be used.

[11] Normality was assessed using the Anderson-Darling statistic (A^2), which tests for non-normality against a null hypothesis of a normal distribution. The unconverted *odds ratio* is marginally non-normal, with the A^2 reporting a *p-value* of 0.09. The *log(odds ratio)* values do not differ appreciably from normality, with A^2 reporting a *p-value* > 0.25. These calculations were performed in *proc univariate* of *SAS*.

Stage Three

The first two stages established that amber is more effective than red, first very roughly in terms of the number of switch pairs and second as a point estimate. Stage Three seeks a more refined estimate of the effectiveness.

The first step is an objective re-assessment of the vehicle pool. Only switch pairs that kept the same wheelbase over the relevant model years are retained. For make-models with multiple switches, this means one switch might be retained with another excluded. In the Nissan Altima (Figure 4 to Figure 7), the first switch from amber (MY 1998-99) to red (MY 2000-2001) is retained, as is the third switch, again amber (MY 2002-2004) to red (MY 2005). However, the second switch from red (MY 2000-2001) to amber (2002-2004) is excluded because the wheelbase changed.[12] It is visually apparent that this was a major redesign, which included reshaping the body and changing the rear lighting.

After the adjustment, twenty-three switch pairs remain. The 2×2 table is constructed for all make-models together. Doing so assumes that the effectiveness of amber is constant and that differences in the effectiveness between make-models are merely chance occurrences. When a vehicle is redesigned, a number of characteristics may change that could influence the likelihood of being involved in an accident – e.g., vehicle handling or stability, driver demographics, and usage patterns – but none of the 23 remaining switch pairs were redesigned to the extent of changing the wheelbase.

Table 4 shows the results. The odds ratio corresponds to an effectiveness of 5.4 percent in favor of amber, and the result is statistically significant ($\chi^2 = 11.59$, with 1 *df*, $p < 0.01$). The 95 percent one-sided confidence interval has a lower bound of 2.8 percent, calculated in SAS™ using the *relrisk* option in *proc freq*.

Table 4: 2×2 table excluding major redesigns

	Striking	Struck	Odds
Red	15,621	15,735	1.007
Amber	14,313	13,632	0.952
			0.946 Odds Ratio
χ^2 11.59			**5.4% Effectiveness**
p value <0.01			≥ 2.9% one-sided CI bound

Stage Four

To further control for changes in vehicle design and/or lighting configuration, photographs of the switch pairs were reviewed and additional switch pairs were excluded if there was some visible change in body style or in the rear lighting configuration. The appendix shows an example of an excluded make-model (Chevrolet Camaro on page 32).

---

[12] A slight adjustment of the relevant model years thus eliminates MY 2002 because it no longer falls into a two-year window when the comparison to MY 2000-2001 is removed.

The complete list of excluded and included make-models is in the appendix (Table 20 on page 31).

After both sets of exclusions (change in wheelbase and/or changes seen in photographs), 11 switch pairs remain.

Table 5 shows the 2×2 table for these switch pairs. The odds ratio corresponds to an effectiveness of 5.3 percent in favor of amber, and the result is statistically significant (χ^2 = 5.17, with 1 df, $p < 0.02$). The 95-percent one-sided confidence interval on the effectiveness extends to 1.5 percent. This result is nearly identical to that of *Stage Three*. However, the comparisons here are more valid because only make-models with body style and rear lighting configuration as identical as possible are included. From a statistical perspective, this result is less precise due to the smaller sample size.

Table 5: 2×2 table excluding minor and major redesigns

	Striking	Struck	Odds	
Red	7,061	7,219	1.022	
Amber	6,848	6,629	0.968	
			0.947	Odds Ratio
χ^2 5.17			**5.3% Effectiveness**
p value 0.01			≥ 1.5% one-sided CI bound

To ensure consistency across make-models, the results of Table 5 can be checked with the methods of Stages One and Two of the analysis. By number, 8 of the 11 favor amber (8 of 11 = 73%), the same as in Stage One but not quite significantly more than 50 percent, owing to the smaller number of switch pairs (11 here versus 33 originally). The mean (4.9%) of the 11 individual effectiveness estimates is similar to the effectiveness from the 2×2 table (5.3%), though the mean is not significantly different from zero. The median of the 11 individual effectiveness estimates is 1.4 percent.

Supplementary Analyses

The estimate from Stage Four of the primary analysis shows a 5.3 percent crash reduction for amber turn signals, compared to red, in crashes where the lead (struck) vehicle is turning left, turning right, changing lanes, merging, parking, or making a U-turn. Several further steps were conducted to examine situations where the effectiveness may differ.

These supplementary analyses were carried out continuing with the dataset of Stage Four, because it is most clear that these switch pairs differ only in rear turn signal color. Objectively, however, the dataset of Stage Three could be used if a larger sample size is desired, on the fact that the results of Stage Three and Stage Four are nearly identical.

Injury Outcome

The first supplementary analysis is conducted on crash injury outcome. The State Data files include a variable at the *crash* level that classifies the most severe injury suffered by any person involved in the crash. The categories vary slightly by State. In the appendix (page 35), there is an example of one State's injury classification, followed by a summary of injury outcome for all States.

Below are the 2×2 tables for crashes with injuries (Table 6) and without injuries (Table 7). The effect of amber turn signals is larger in injury crashes (8.3%, with a 95% confidence interval extending to 1.3%), compared to non-injury crashes (2.8%, confidence interval extending beyond zero). On one hand, the estimates might be considered equal because the confidence intervals largely overlap. On the other hand, the effect is significantly different from zero for injury crashes but not for non-injury crashes.

Table 6: 2×2 table for crashes with injuries

	Striking	Struck	Odds
Red	2,049	2,079	1.015
Amber	1,956	1,819	0.930

χ^2 3.74
p value 0.03

0.917 Odds Ratio
8.3% Effectiveness
≥ 1.3% one-sided CI bound

Table 7: 2×2 table for crashes without injuries

	Striking	Struck	Odds
Red	4,707	4,831	1.026
Amber	4,563	4,551	0.997

χ^2 0.96
p value 0.16

0.972 Odds Ratio
2.8% Effectiveness
(not significant) one-sided CI bound

Vehicle Maneuver

The tables on the following page investigate the effectiveness of amber turn signals according to the maneuver of the struck vehicle. The effectiveness estimates are 6.4 percent for changing lanes or merging (Table 8), 4.0 percent for all secondary maneuvers (Table 9), 4.2 percent for right turns (Table 10), and 5.6 percent for left turns (Table 11). Only the effectiveness for left turns is large enough in magnitude and sample size to be significantly different from zero ($\chi^2 = 3.05$, $df = 1$, $p < 0.05$, with a 95 percent confidence interval extending to 0.3%). None of the four can be said to be significantly different from the others, implying that amber turn signals are equally effective in all types of rear impacts with assumed signal engagement.

Table 8: 2×2 table for crashes where the struck vehicle was changing lanes or merging

	Striking	Struck	Odds
Red	843	798	0.947
Amber	966	856	0.886
			0.936 Odds Ratio

χ^2 0.94   **6.4% Effectiveness**
p value 0.17   (not significant) one-sided CI bound

Table 9: 2×2 table for crashes where the struck vehicle was changing lanes, merging, parking, or making a U-turn

	Striking	Struck	Odds
Red	1,095	1,017	0.929
Amber	1,259	1,123	0.892
			0.960 Odds Ratio

χ^2 0.46   **4.0% Effectiveness**
p value 0.25   (not significant) one-sided CI bound

Table 10: 2×2 table for crashes where the struck vehicle was turning right

	Striking	Struck	Odds
Red	2,196	2,314	1.054
Amber	2,053	2,072	1.009
			0.958 Odds Ratio

χ^2 1.00   **4.2% Effectiveness**
p value 0.16   (not significant) one-sided CI bound

Table 11: 2×2 table for crashes where the struck vehicle was turning left

	Striking	Struck	Odds
Red	3,747	3,859	1.030
Amber	3,511	3,412	0.972
			0.944 Odds Ratio

χ^2 3.05   **5.6% Effectiveness**
p value 0.04   ≥ 0.3% one-sided CI bound

Gender

Table 12 (males) and Table 13 (females) continue the analysis based on the gender of the driver. The effectiveness estimates are nearly identical: 5.5 percent for males and 6.0 percent for females. In fact, these tables highlight the utility of comparisons based on the odds ratio with striking vehicles as the exposure metric. In general, males are more often the driver of the striking vehicle than females, and less often the driver of the struck vehicle. However, the differences in odds wash out, making effectiveness equal.

Table 12: 2×2 table for crashes where the driver was male

	Striking	Struck	Odds
Red	3,622	3,214	0.887
Amber	3,507	2,941	0.839

0.945 Odds Ratio

χ^2 2.63   **5.5% Effectiveness**

p value 0.05   (not significant) one-sided CI bound

Table 13: 2×2 table for crashes where the driver was female

	Striking	Struck	Odds
Red	3,226	3,863	1.197
Amber	3,151	3,548	1.126

0.940 Odds Ratio

χ^2 3.24   **6.0% Effectiveness**

p value 0.04   ≥ 0.5% one-sided CI bound

Vehicle Age

The age of the vehicle at the time of the crash was analyzed in NHTSA reports of other lighting regulations. It was found that the likelihood of being rear-struck decreases with increasing vehicle age. In the current analysis, vehicle age could be influential if the relative proportion of red to amber vehicles differs with vehicle age. For example, if there are relatively more "old" amber vehicles, the observed effects in favor of amber might in part be a result of being older rather than being amber.

It may occur that effectiveness is a function of vehicle age in itself, due to something like dulling or dirtying of the lenses that affected one color more than the other. Age also approximates the role of vehicle miles traveled – if there was some reason why the rate of being rear-struck compared to being rear-impacted varied with miles traveled.

The nature of this evaluation differs from other NHTSA reports on automotive lighting. With CHMSL, for example, it was true that (nearly) every older vehicle lacked the equipment and that every newer vehicle was equipped, thus the change was unidirectional. This is not true for turn signal color, where there are both amber-to-red and red-to-amber switches. Further, most of the switches overlap the years for which data is available. Therefore, it is possible (in most cases) for each make-model to have crash involvements for both red and amber at the same age (e.g., a Ford Taurus with amber signals can have a crash at age 5, and a Ford Taurus with red signals can have a crash at age 5).

Age at the time of crash is calculated as *Calendar Year* <minus> *Model Year* <plus> *One*. For example, a 1996 MY Ford Taurus in a crash in 2000 would be assigned an age of 5 (2000 − 1996 + 1 = 4 + 1 = 5)[13].

For each make-model, age is restricted as a result of State data being available from 1989 to 2005 (2006 in Michigan), with some gaps as noted in Figure 8. The earliest make-models, with MY prior to 1989, can never be involved in crashes as "young" vehicles. The Chevrolet Celebrity, for example, must be at least 5 years old – its latest MY is 1985, with the earliest crash possible in 1989 (1989 − 1985 + 1 = 5). Thus age 5 and 6 for the Celebrity are attainable only for the red version, because the amber version will need to be at least 7 years old, based on the amber version in MY 1983.

The more recent make-models are restricted at the upper end, where the earliest MY will be able to "achieve" crashes at an older age, up to the limit of accidents in calendar year 2005. Therefore, the "oldest" block for each make-model will be monochromatic. The MY 1996 Ford Taurus (red) can achieve a maximum age of 10 years (2005 − 1996 + 1 = 10), compared to only 8 years for the first year of the amber version (MY 1998).

The number of struck vehicles is shown in Table 14, covering all ages from 0 to 24. For perspective, the ages 1 to 10 comprise 85 percent of the total cases. The column *Red Excess* is constructed to represent the percent more red (e.g., at age 3, 741 red ÷ 735 amber = 1.01 <minus> one = 0.01 = +1%). There is an excess of red vehicles at most ages, and the percent excess is in single digits up to 27 percent at the early ages that constitute the bulk of the data. From age 10 onwards, this particular mix of make-models is relatively more populated by vehicles with red signals. Age 1 to 10 comprise 84 percent of the data, and the excess of red on this range is only +3 percent.

Compared to the overall excess (bottom row) of +9 percent red, the younger ages are more equally-populated by amber – there are relatively more old red vehicles. Thus, if older vehicles are *less likely* to be rear-struck, the odds of the red vehicles in this study being rear-struck should be slightly lower than if the ratio of amber to red vehicles were constant across vehicle age. In turn, the effectiveness of amber would be slightly lower. In a sense, this means the study should *not* be biased in favor of amber, as a function of vehicle age. On the other hand, the true effectiveness of amber may be slightly higher than the reported +5.3 percent.

---

[13] The "plus one" accounts for vehicles sold in the calendar year before the model year. Thus a "brand new" vehicle, involved in a crash during the calendar year preceding its model year, will have an age of zero.

Table 14: Number of struck vehicles by age at time of crash

Age	Amber	Red	Red Excess
0	24	58	+142%
1	532	505	-5%
2	733	686	-6%
3	735	741	+1%
4	682	666	-2%
5	690	875	+27%
6	691	797	+15%
7	664	671	+1%
8	506	431	-15%
9	314	286	-9%
10	180	238	+32%
11	105	193	+84%
12	81	203	+151%
13	140	172	+23%
14	129	176	+36%
15	119	144	+21%
16	84	103	+23%
17	74	86	+16%
18	52	70	+35%
19	44	47	+7%
20	26	35	+35%
21	18	26	+44%
22	5	5	0%
23	1	4	+300%
24	0	1	--
Total	6,629	7,219	+9%

Figure 2 shows the odds ratio for vehicles age 1 to 10. The dashed line displays the combined odds ratio on this range, which is nearly identical to the overall odds ratio. In this region, the confidence intervals extend less than 0.25. There is no clear pattern to the data, thus any model-fitting would be overkill at this stage.

Figure 2: Odds ratio by vehicle age, with 95 percent one-sided confidence intervals (reference marked for overall odds ratio)

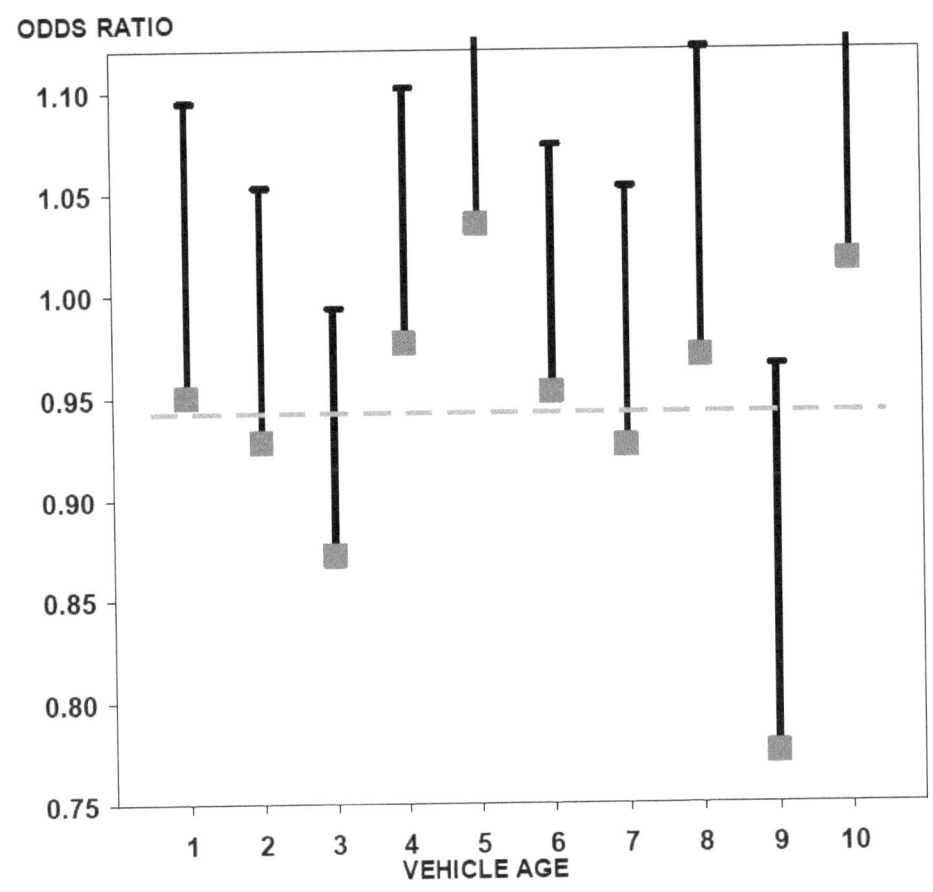

Comparisons can be made at the level of make-models – creating the 2×2 table for each age of each make-model, e.g., comparing 5-year-old Ford Rangers with red turn signals to 5-year old Ford Rangers with amber turn signals. Due to restrictions in the possible ages, there are one hundred twenty-one 2×2 tables with all non-zero elements. Many tables have small cell counts, and only the most common make-models accumulate cell counts above 100. From the 121 make-model-ages, 110 comparisons can be made from age n to $n+1$ (note: 121 <minus> 11 = 110). For example, 1-year-old Ford Tauruses have an effectiveness of 1 percent, compared to 10 percent for 2-year-old Ford Tauruses. Thus, one year of vehicle age corresponds to an increase in effectiveness. Across all 110 comparisons, there were 55 increases in effectiveness and 55 decreases in effectiveness – *exactly half*. The split is near 50/50 if restrictions are introduced to ignore comparisons with small sample sizes – it does not matter what exact criterion is used for exclusion.

Observed effectiveness, State by State

The effectiveness of amber turn signals is calculated within each of the 14 States. Figure 3 shows the odds ratios for each State, with the 95 percent one-sided confidence intervals marked. Ten of the 14 have odds ratios below 1 – favoring amber. The median effect is a

5.2 percent percent reduction for amber, close to the basic result with pooled data (5.3%, Table 5). The small sample sizes give rise to wide confidence intervals, resulting in only one State (Texas) where amber is significantly more effective than red. This Texas result is curious because amber is *so much more* effective than in the other States. Despite the use of its own make-model coding (in lieu of VIN's), 10 of the 11 switch pairs are present. This result may be a chance occurrence or else due the intersection of several small factors favoring amber that are in themselves minor (e.g., injury outcome, Table 6). There is no obvious reason why one state would be so different.

Figure 3: Odds ratio by State, with 95 percent one-sided confidence intervals (reference marked for 1.00)

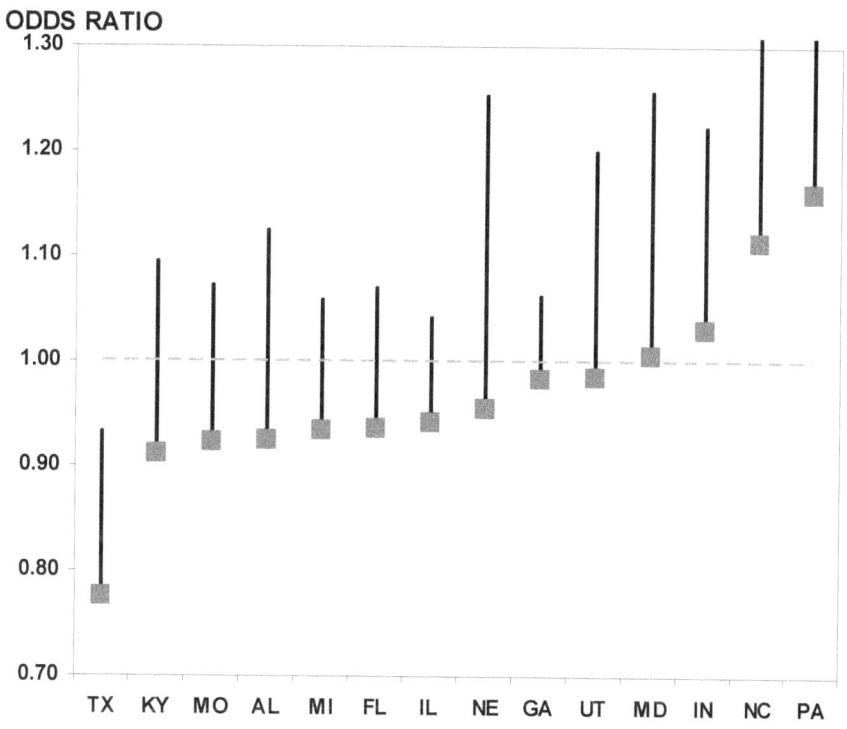

*Alternate Control Group: Rear impacts where the struck vehicle was **not** turning, changing lanes, merging, parking, or making a U-turn*

Table 15 presents the results when the control group consists of rear impacts where the struck vehicle was not turning, changing lanes, merging, parking, or making a U-turn. For example, the struck vehicle might have been going straight ahead, slowing or stopped, and generally less likely to be engaging turn signals. The crashes of interest remain those where signal engagement is assumed – rear-struck while turning, changing lanes, merging, or parking (*cf.* Table 5).

Table 15: 2×2 table using non-turning rear-struck as the control group

	Striking	Struck	Odds
Red	52,869	7,219	0.137
Amber	51,710	6,629	0.128

0.939 Odds Ratio
6.1% Effectiveness
≥ 3.3% one-sided CI bound

χ^2 12.15
p value <0.01

The effectiveness in favor of amber is 6.1 percent. The result is statistically significant different from zero (χ^2 = 12.15 with 1 *df*, $p < 0.01$). The 95 percent confidence interval extends to 3.3 percent. The result is slightly stronger in favor of amber compared to the primary analyses, although the confidence intervals are largely overlapping. When analyzed within State, the results are comparable to Figure 3, and the median effectiveness of the 14 is 6.8 percent. As discussed in the selection of the primary control group, this alternate control group is likely imprecise because there may be a large number of vehicles that are signaling but not engaged in a maneuver that serves to identify their intention as such (e.g., waiting at a stop sign to turn right).

Appendix

The appendix contains information for readers who are interested in the technical details of the data sources and methods. The relative placement within the body of the report is identified by the section headings.

Methods: Turn Signal Color Identification

The figures to follow show the rear-lighting color and configuration for the Nissan Altima. This was the only make-model found to have switched color three times across a small time frame, being initially amber (MY 1998-1999, Figure 4) and switching to red (MY 2000-2001, Figure 5). A reversion to amber (MY 2002-2004, Figure 6) corresponded with a change in body style and wheelbase, as well as reconfiguring the rear lighting assembly. This redesigned Altima then switched to red (MY 2005, Figure 7).

Figure 4: Nissan Altima, MY 1998-1999, amber turn signals

Figure 5: Nissan Altima, MY 2000-2001, red turn signals

Figure 6: Nissan Altima, MY 2002-2004, amber turn signals

Figure 7: Nissan Altima, MY 2005, red turn signals

Methods: Data Sources

Figure 8 shows the States and the available calendar years of data that meet the inclusion criteria listed on page 9.

Figure 8: States and calendar years of State data included in study

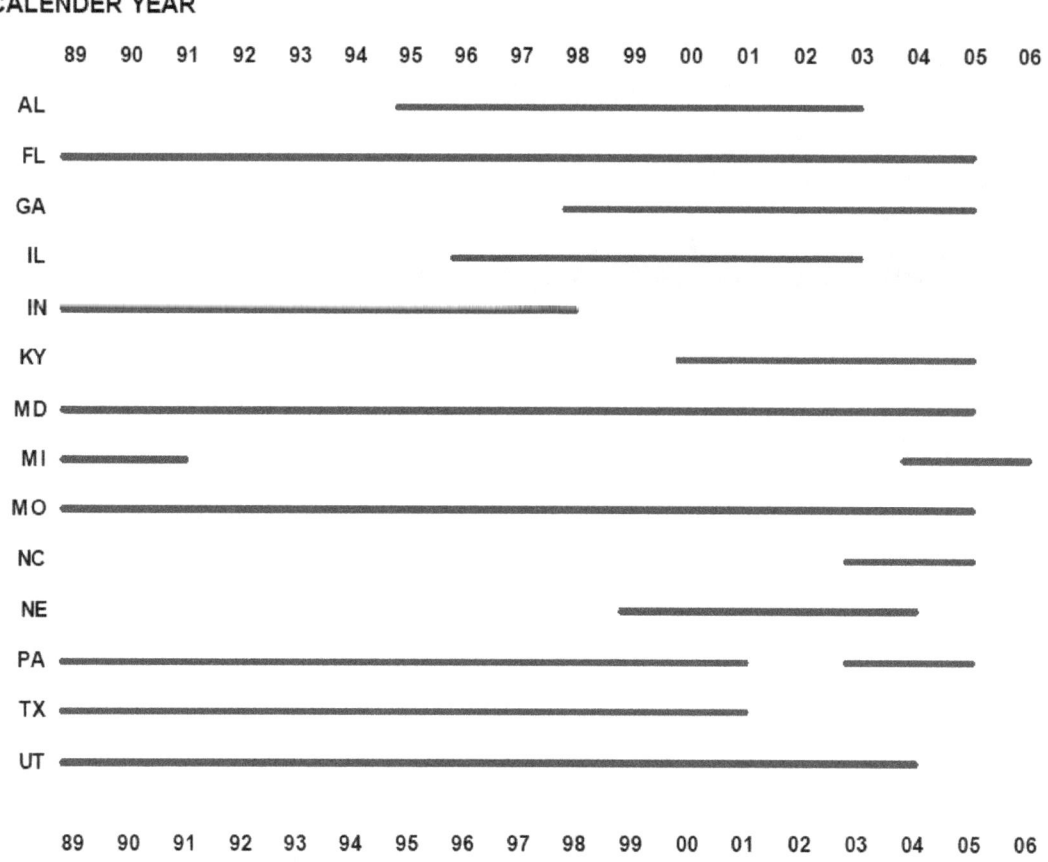

Table 16 shows the types of crashes identifiable for each of the 14 States used in this study. The left-hand panel identifies the secondary types of crashes – changing lanes, merging, parking, un-parking, and U-turn. Even when available, these categories are not necessarily identifiable for all calendar years of data. The right-hand panel shows that every State has crashes identified as turning right and turning left. Two States, North Carolina and Texas, contain extra vehicle maneuvers that clearly identify crashes where the lead vehicles were turning but do not classify the turns as left or right. In several States, there is no distinction between entering parking and leaving parking, and these are lumped into the column "Parking." On the other hand, three States specify leaving a parking space or a parked position (labeled here "Un-Parking"). Only one State, North Carolina, identifies both parking and un-parking.

Table 16: Summary of types of crashes, by State

	Ch. Lane	Merging	Parking	Un-Parking	U-Turn	Right Turn	Left Turn	Turning
AL	X	X	X		X	X	X	
FL	X		X		X	X	X	
GA	X		X		X	X	X	
IL	X	X			X	X	X	
IN	X	X	X		X	X	X	
KY	X	X	X		X	X	X	
MD	X		X		X	X	X	
MI	X		X		X	X	X	
MO	X			X	X	X	X	
NC	X		X	X	X	X	X	X
NE	X	X		X	X	X	X	
PA	X	X	X		X	X	X	
TX	X				X	X	X	X
UT	X	X			X	X	X	

Methods: State Data Summary

This section summarizes the dataset based on the make-models and model years used in the study. Figure 9 shows the number of struck vehicles in terms of left turn, right turn, and all other maneuvers; Figure 10 contains the same numbers on a percentage basis, by State.

Table 17 shows the number of struck vehicles for all relevant maneuvers. These vehicles are most commonly rear-struck while turning left. The number of left-turning vehicles is generally around two-to-three times greater than the number of right-turning vehicles. The exceptions are Indiana and Missouri, where the number of right-turning vehicles is greater than the number of left-turning vehicles. In sum, there are 63 percent more left-turning than right-turning struck vehicles. The number of non-turning (i.e., changing lanes, parking, un-parking, merging, or making a U-turn) struck vehicles varies from a little over 20 percent to around 5 percent. Because Georgia has a relatively high percentage of non-turning cases and it is the State with the most data, any analysis limited to these cases will be especially influenced by this one State.

Figure 9: Number of struck vehicles, by vehicle maneuver for 14 States

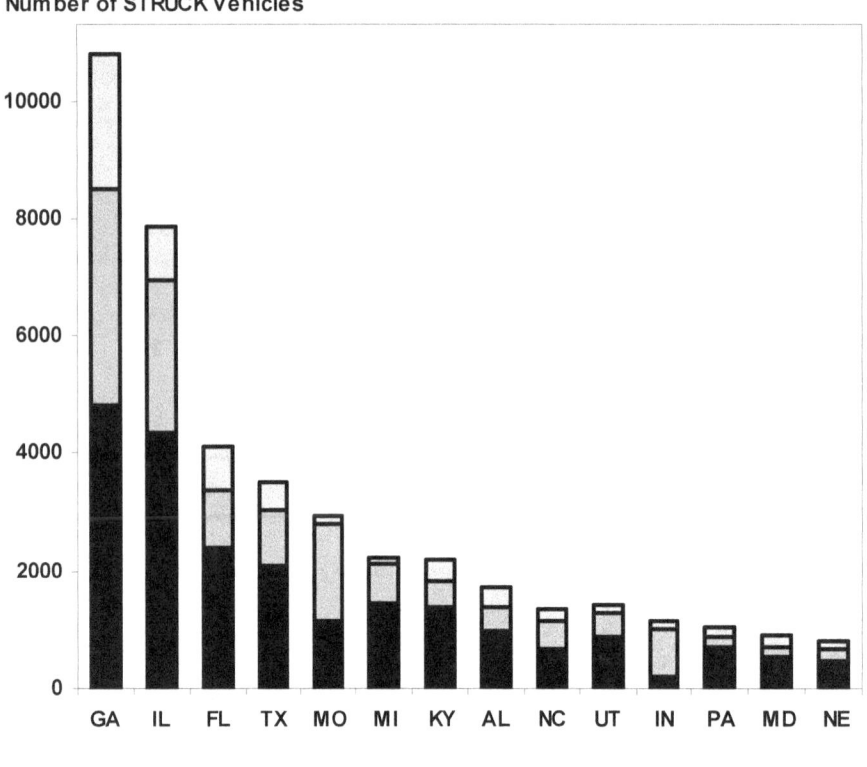

Figure 10: Relative frequency of struck vehicle maneuvers, by State

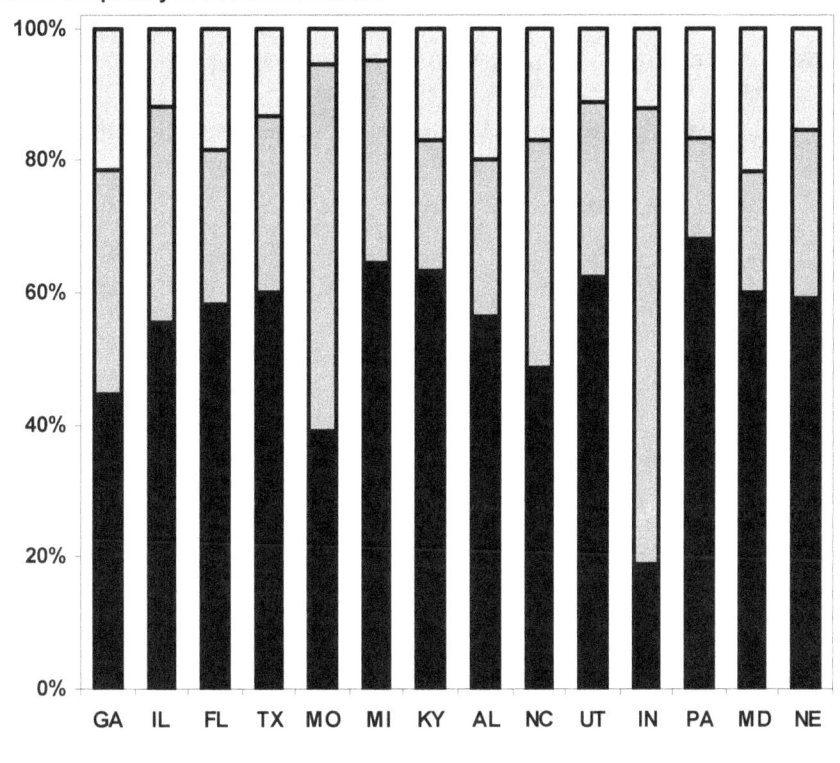

Table 17: Counts of collision types for struck vehicles, by State

	Ch. Lane	Merging	Parking	Un-Parking	U-Turn	Right Turn	Left Turn	Turning	TOTAL
GA	1,542	-	604	-	160	3,685	4,817	-	10,808
IL	582	250	-	-	88	2,586	4,360	-	7,866
FL	569	-	36	-	155	957	2,403	-	4,120
TX	453	-	-	-	18	934	2,105	1	3,511
MO	128	-	-	12	19	1,639	1,152	-	2,950
MI	89	-	4	-	14	684	1,434	-	2,225
KY	184	135	31	-	17	438	1,381	-	2,186
AL	170	140	3	-	28	409	969	-	1,719
NC	149	-	14	38	28	475	662	183	1,549
UT	96	14	-	-	48	380	889	-	1,427
IN	78	40	10	-	12	797	218	-	1,155
PA	79	22	35	-	36	157	704	-	1,033
MD	127	-	17	-	52	166	545	-	907
NE	89	20	-	7	10	209	482	-	817
TOTAL	4,335	621	754	57	685	13,516	22,121	184	
									42,273

Figure 11 and Figure 12 show the same breakdown of vehicle maneuvers for each make-model. The labeling on the figures is listed with Table 18, immediately afterwards. The counts are combined across all model years, with no distinction based on turn signal color at this stage. The purpose is to give a sense of the scale, rather than to imply that certain make-models are more apt to be rear-struck in the situations depicted in Table 17. The number of crash involvements is a function of many factors, primarily the number of vehicles registered in each State during the calendar years used in this analysis. There is greater variability in vehicle maneuver by State than by make-model.

Figure 11: Number of struck vehicles, for the 26 make-models that switched signal color

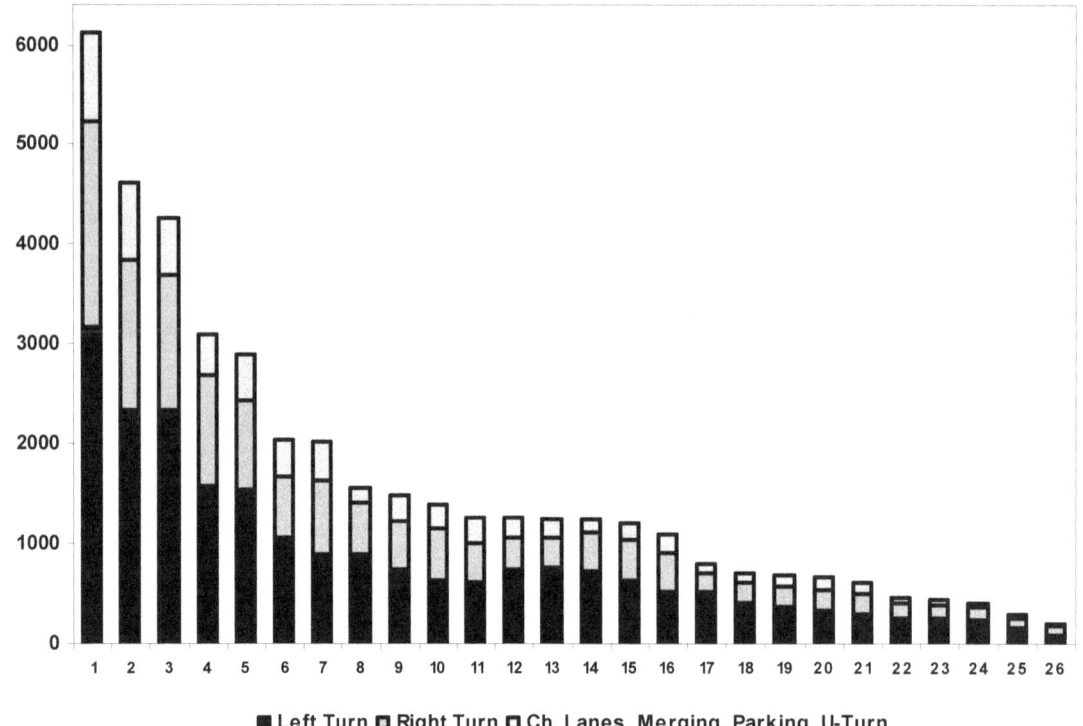

Figure 12: Relative frequency of struck vehicle maneuvers, by make-model

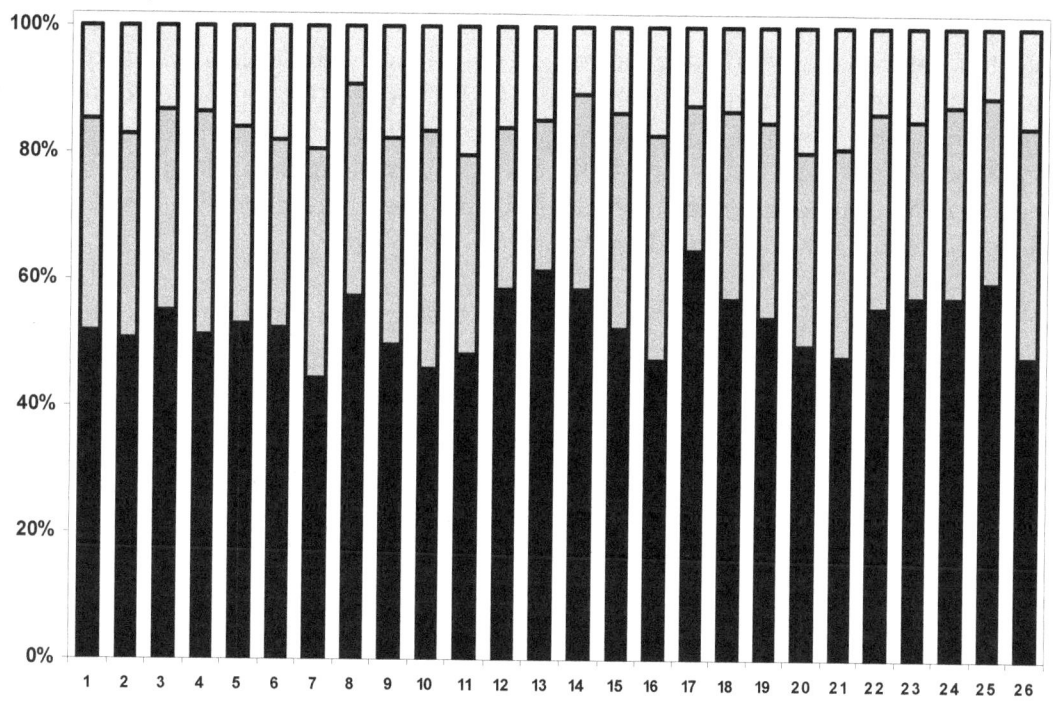

Table 18: Counts of collision types for struck vehicles, by make-model

	Ch. Lanes	Merging	Parking	Un-Parking	U-Turn	Right Turn	Left Turn	Turning	TOTAL
1 Ford Taurus(4)	584	111	100	6	105	2,048	3,176	21	6,151
2 Ford Explorer	544	55	121	6	62	1,506	2,333	31	4,658
3 Pontiac GrandAm	362	89	43	6	67	1,349	2,346	14	4,276
4 Ford Escort (noSW)	284	34	49	5	49	1,098	1,583	12	3,114
5 Toyota Corolla	304	48	56	1	51	903	1,536	11	2,910
6 Chevrolet Camaro	235	46	32	3	48	604	1,064	10	2,042
7 Nissan Altima	284	26	40	4	35	729	895	17	2,030
8 Chevrolet Celebrity	98	16	13	-	17	517	890	-	1,551
9 Honda Civic(coupé)	185	17	30	1	32	484	739	5	1,493
10 Honda Civic(sedan)	159	22	28	4	17	520	637	10	1,397
11 Tahoe / Yukon	173	15	37	9	23	398	613	7	1,275
12 Mustang(90s)	144	13	19	1	26	321	739	5	1,268
13 Olds 88	112	16	30	2	24	294	764	1	1,243
14 Buick Century	89	13	13	-	16	383	726	1	1,241
15 Dodge Intrepid	106	25	18	1	14	411	628	5	1,208
16 Ford Ranger	138	6	29	1	15	390	518	4	1,101
17 Chevrolet Nova	67	4	10	1	17	181	515	-	795
18 Mustang(80s)	69	6	6	-	13	206	400	-	700
19 Dodge Stratus(4d)	67	15	12	-	8	211	368	7	688
20 Suburban	90	4	21	3	13	202	327	4	664
21 Honda Accord(2d)	82	11	19	2	4	204	297	6	625
22 Ford Taurus(SW)	41	6	7	-	9	143	256	7	469
23 Pontiac Firebird	47	8	5	-	6	126	258	-	450
24 Pontiac GP (coupé)	31	9	5	-	7	125	238	2	417
25 Chrysler LeBaron	22	4	5	-	1	85	173	2	292
26 Saturn LS	18	2	6	1	6	78	102	2	215
TOTAL	4,335	621	754	57	685	13,516	22,121	184	42,273

Table 19 shows the number of struck vehicles relative to the number striking and all other crash types. The numbers of struck and striking are nearly identical, as they should be based on the crash definitions. Across the 14 States, the struck vehicles comprise 2.1 percent of the total crashes (defined as Struck + Striking + All Others). The percentage struck varies from 0.8 percent (Pennsylvania) to 3.9 percent (Georgia). Ten of the 14 fall within one percentage point of the overall percentage. The differences arise primarily from reporting and inclusion criteria in the State Data System, rather than to imply that drivers in certain states are much more apt to be involved in rear-end crashes. The impact of these differences is examined after the primary analyses.

Table 19: Struck and Striking cases relative to overall number of crashes, by State

	Struck	Striking	All Others	% Struck
GA	10,808	11,187	257,478	3.9%
IL	7,866	7,659	237,613	3.1%
UT	1,427	1,348	45,876	2.9%
NE	817	816	30,717	2.5%
IN	1,155	1,086	47,504	2.3%
KY	2,186	2,152	93,149	2.2%
MI	2,225	2,167	125,304	1.7%
NC	1,549	1,564	94,117	1.6%
AL	1,719	1,774	105,211	1.6%
FL	4,120	4,355	253,339	1.6%
MO	2,950	2,665	190,099	1.5%
TX	3,511	3,336	233,859	1.5%
MD	907	885	102,061	0.9%
PA	1,033	1,067	128,383	0.8%
TOTAL	**42,273**	**42,061**	**1,944,710**	**2.1%**

Results: Stage One

Table 20 shows the effectiveness of amber signals relative to red signals for each of the 33 switch pairs used in this analysis. The two stages of exclusions are noted as (1) body-size changes and (2) styling changes. The effectiveness calculations vary widely. In order for an effectiveness to be significantly different from zero, both a large sample size and high effectiveness are necessary. For example, the Ford Taurus sedan with an effectiveness of +25 percent is significantly different from zero, whereas the Ford Ranger with an effectiveness of +17 percent is not.

Table 20: List of effectiveness for 33 make-models

	Change	Effectiveness	Amber MY	Red MY	Exclusion
Chevrolet Camaro	Amber to Red	+33%	1991 - 92	1993 - 94	
Honda Accord (2d)	Amber to Red	+31%	1996 - 97	1998 - 99	1
Pontiac GrandAm	Amber to Red	+26%	1998 - 98	1999 - 00	1
Ford Taurus (4d)	Red to Amber	+25%	1996 - 97	1994 - 95	
Ford Explorer	Amber to Red	+21%	2000 - 01	2002 - 03	1
Pontiac Firebird	Amber to Red	+18%	1991 - 92	1993 - 94	2
Chevrolet Celebrity	Amber to Red	+18%	1982 - 83	1984 - 85	
Ford Ranger	Amber to Red	+17%	1998 - 99	2000 - 01	
Ford Taurus (SW)	Red to Amber	+16%	1996 - 97	1994 - 95	
Nissan Altima	Amber to Red	+16%	1998 - 99	2000 - 01	
Nissan Altima	Red to Amber	+15%	2002 - 03	2000 - 01	1
Nissan Altima	Amber to Red	+12%	2003 - 04	2005 - 05	
Mustang (80s)	Red to Amber	+11%	1983 - 84	1982 - 82	2
Ford Escort (noSW)	Amber to Red	+11%	1995 - 96	1997 - 98	2
Dodge Stratus (4d)	Amber to Red	+9%	1999 - 00	2001 - 02	2
Honda Civic (sedan)	Amber to Red	+9%	1999 - 00	2001 - 02	2
Buick Century	Amber to Red	+8%	1982 - 83	1984 - 85	
Toyota Corolla	Red to Amber	+8%	1993 - 94	1991 - 92	1
Ford Explorer	Red to Amber	+7%	1995 - 96	1993 - 94	1
Saturn LS	Amber to Red	+3%	2001 - 02	2003 - 04	2
Olds 88	Red to Amber	+1%	1984 - 85	1982 - 83	
Ford Taurus (4d)	Amber to Red	+1%	1996 - 97	1998 - 99	1
Tahoe/Yukon	Red to Amber	+1%	2000 - 01	1998 - 99	
Pontiac GP (coupé)	Amber to Red	+0%	1995 - 96	1997 - 98	1
Honda Civic (coupé)	Amber to Red	-3%	1997 - 98	1999 - 00	2
Chevrolet Nova	Amber to Red	-3%	1986 - 86	1987 - 88	
Pontiac GrandAm	Red to Amber	-7%	1996 - 97	1994 - 95	2
Chevrolet Camaro	Red to Amber	-11%	1997 - 98	1995 - 96	2
Suburban	Red to Amber	-14%	2000 - 01	1998 - 99	1
Mustang (90s)	Amber to Red	-23%	1992 - 93	1994 - 95	1
Ford Taurus (SW)	Amber to Red	-29%	1996 - 97	1998 - 99	1
Dodge Intrepid	Red to Amber	-30%	1998 - 99	1996 - 97	2
Chrysler LeBaron	Red to Amber	-36%	1993 - 94	1991 - 92	2

Results: Stage Four

An example of the Chevrolet Camaro is shown in Figure 13 (MY 1991-1992) and Figure 14 (MY 1993-1994). That switch pair is now excluded because of the noticeable change in the rear lights from a rectangular to a somewhat smaller, oval configuration. By contrast, the later switch pair for the Camaro is retained (Figure 15, MY 1995-1996; Figure 16, MY 1997-1998) because the rear lighting is of the same size and shape. At this level of detail, it is not possible to be entirely sure of the rear lighting configuration. For the Camaro MY 1993-1996 with red turn signals, it is not clear if the turn signal is housed separately from the brake lamp or if there is a single lamp with shared function.

This introduces a confounding factor if the effect of being separated is important, as suggested by laboratory studies. Other changes that affect vehicle dynamics were not evaluated, e.g., changes to the brake and/or steering systems.

Figure 13: Chevrolet Camaro, MY 1991-1992, amber

Figure 14: Chevrolet Camaro, MY 1993-1994, red

Figure 15: Chevrolet Camaro, MY 1995-1996, red

Figure 16: Chevrolet Camaro, MY 1997-1998, amber

Two tables follow that summarize the 33 original switch pairs and note the exclusions. They are based on the direction of the switch: red to amber in Table 21 and amber to red in Table 22. Make-models that switched more than once are listed in both tables, as appropriate based on the direction of the switch. Those excluded in the first stage are noted simply as "Wheelbase" under the *Exclusion* column. The second set of exclusions is noted as a body style modification or re-configuration of the rear lighting housing. The *Comment* column notes details, such as relevant body types, to specify the applicable make-model. In some cases, only a certain body type for a make-model (e.g., the Chrysler LeBaron convertible) was identified as having changed signal color.

Table 21: Make-models that switched from red To amber (13)

Make-Model	Red MY	Amber MY	Comment	Exclusion
Olds 88	1982-83	1984-85		
Ford Mustang	1982	1983-84	Extent of multiple stylings not apparent	Re-configured lighting
Toyota Corolla	1991-92	1993-94		Wheelbase
Chrysler LeBaron	1991-92	1993-94	Convertible	Re-configured lighting
Ford Explorer	1993-94	1995-96		Wheelbase
Ford Taurus Sedan	1994-95	1996-97		Wheelbase
Ford Taurus Station Wagon	1994-95	1996-97		Wheelbase
Pontiac GrandAm	1994-95	1996-97		Re-configured lighting
Chevrolet Camaro	1995-96	1997-98		
Dodge Intrepid	1995-96	1997-98		Red have tinted housing
GMC Tahoe/Yukon	1998-99	2000-01		
Chevrolet Suburban	1998-99	2000-01		Wheelbase
Nissan Altima	2000-01	2002-03		Wheelbase

Table 22: Make-models that switched from amber To red (20)

Make-Model	Amber MY	Red MY	Comment	Exclusion
Chevrolet Celebrity	1982-83	1984-85		
Buick Century	1982-83	1984-85		
Chevrolet Nova	1986	1987-88	1985 is pre-CHMSL	
Chevrolet Camaro	1991-92	1993-94		Body style
Pontiac Firebird	1991-92	1993-94		Body style
Ford Mustang	1992-93	1994-95	Exclude convertibles	Wheelbase
Ford Escort	1995-96	1997-98	Exclude station wagons	Body style
Pontiac GP	1995-96	1997-98	Only coupé[14]	Wheelbase
Honda Accord	1996-97	1998-99	Only coupé	Wheelbase
Ford Taurus Sedan	1996-97	1998-99		
Ford Taurus Station Wagon	1996-97	1998-99		
Pontiac GrandAm	1997-98	1999-2000		Wheelbase
Honda Civic	1997-98	1999-2000	Only coupé	Rear lighting re-configured
Ford Ranger	1998-99	2000-01		
Nissan Altima	1998-99	2000-01		
Honda Civic	1999-2000	2001-02	Only sedan	Body style
Dodge Stratus	1999-2000	2001-02	Only sedan	Body style
Ford Explorer	2000-01	2002-03		Wheelbase
Saturn LS	2001-02	2003-04	Wagon (LW) too uncommon for inclusion but also changed	Body style and rear light housing
Nissan Altima	2003-04	2005-06		

Supplementary Analyses: Injury Outcome

Table 23 shows the values for Florida, which are typical. The three stages for injuries are grouped because not all States distinguish injury severity, e.g., Illinois classifies crashes as only "fatal," "injury," and "property damage only."

---

[14] The European spelling *coupé* is sometimes encountered in North American English, e.g., in *The Great Gatsby*. In this case, it is appropriate because European motor vehicle standards mandate amber rear turn signals.

Table 23: Crash Severity for Florida

Description from State Data Manual	Class
Unknown (2002–Present)	*Unknown*
No injury—indicates there is no reason to believe any person received bodily harm from the motor vehicle crash	*None*
Possible injury—no visible signs of injury but complaint of pain or momentary unconsciousness	*Injury*
Non-incapacitating injury—any visible injuries such as bruises, abrasions, limping, etc.	*Injury*
Incapacitating injury—any visible signs of injury from the crash and person(s) had to be carried from the scene	*Injury*
Fatal injury—any injury sustained in the motor vehicle crash that results in death within 90 days	*Fatality*

Table 24 shows the number of injury types for the struck vehicles in the dataset with wheelbase and styling changes removed. The number of fatalities is, fortunately, quite small and certainly insufficient to include as a classification. The number of unknown classifications is generally small, except for Georgia. Pennsylvania and especially Nebraska have high percentages of unknown cases but they are few compared to Georgia in absolute number. The percent injuries (excluding fatalities and others) varies considerably, from 65 percent (Texas) to 15 percent (Alabama). This arises from differences in case inclusion (reporting threshold) for the various States, rather than to suggest that crashes more commonly lead to injury in Texas compared to Alabama.

Table 24: Classification of crashes by injury, struck vehicles only

State	Fatalities	Injuries	Non-Injuries	Other	Percent Injuries	Total Cases
TX	0	601	330	0	65%	931
FL	2	717	466	0	61%	1,185
PA	0	167	146	25	53%	338
MD	0	173	167	0	51%	340
NE	0	73	126	48	37%	247
NC	2	130	320	4	29%	456
MI	1	403	1,076	0	27%	1,480
UT	0	132	361	0	27%	493
MO	0	228	756	0	23%	984
IL	1	426	1772	0	19%	2,199
IN	0	129	540	0	19%	669
GA	1	532	2,305	484	19%	3,322
KY	0	106	545	0	16%	651
AL	0	81	472	0	15%	553
TOTAL	7	3,898	9,382	561	29%	13,848

www.ingramcontent.com/pod-product-compliance
Lightning Source LLC
Chambersburg PA
CBHW081400170526
45166CB00010B/3156